Researching French Local History and Geneology

INTRODUCTION

This selected bibliography focuses on published books at the Library of Congress. It concentrates on books that describe the records, the locations of records, and the research strategies to conduct French genealogy at the Library of Congress. There are also a few important websites included.

The bibliography is the first of two parts: general books about French genealogy considered as a national topic (including the genealogy at national institutions, or of all the departments and provinces), comprised of the following formats: genealogical handbooks (with research strategies and descriptions of records, libraries, and archives), archival resources (describing archival records, in general, and the archives and libraries in which they are deposited), bibliographies, biographical resources, local history and geographic resources (gazetteers and local history), dictionaries of genealogical terms, and French surnames.

The second part is forthcoming and will list French geographic departments and provinces, under which are listed many of the same topics listed under general books, but for a smaller, more specific area.

In general, the entries are in alphabetical order by main entry. However, there is an exception under topics such as "Genealogy" if the book is a general handbook; and under "Archives" if the book is a general handbook to the region, it is listed first.

I.   GENERAL FRANCE:

A.   Genealogical Handbooks

French genealogical handbooks describe how to use French records to compile a genealogy. They describe the kinds of records available for genealogical research, and where to find them. Because genealogical records are dispersed geographically in France, it is necessary for the researcher to know the place of origin of the family being researched. There are only a few English language handbooks about French genealogy: these are listed first, followed by a selection of more recent handbooks written in French. Handbooks which describe records in smaller geographic divisions (usually provinces or departments) are described under those geographic entities

1.   English

Audin, Margaret.
Barking up that French tree. Owensboro, Ky.: Cook-McDowell Publications, 1980. 61 p., ill., bibliography.
CALL NUMBER: CS599 .A93 1980 <LH&G>
    Includes a case study, copies of original documents, Republican calendar conversion tables, and a glossary of French terms for genealogists. Bibliography, p. 56.
81101710

Baxter, Angus.
In search of your Canadian roots. 3rd ed.. Baltimore: Genealogical Pub. Co., 2000. xxi, 376 p. CALL NUMBER: CS82 .B389 2000<LH&G>
      Because so many Canadians trace their ancestry back to France, there is a short chapter, in English, about French genealogy. Descriptions of civil registration (vital records), Catholic church records, Protestant and Huguenot records, wills, censuses, emigration records, internal passports, voters' lists, conscription lists, military and naval records, passenger lists, indentures, legal records, tax records and cemetery inscriptions. Includes a list of departmental archives. Includes bibliographical references (p. 365-371) and index.
99073935

Durye, Pierre.
Genealogy: an introduction to Continental concepts. Translated from the French language by Wilson Ober Clough. New Orleans: Polyanthus, 1977. 143 p.:ill. bibliography.
CALL NUMBER: CS17 .D813 <LH&G>
      Translation of *La généalogie*, 4th ed. English version translated from 1975 edition first published in Paris in 1961 by Presses Universitaires de France. Discusses the historical background of French genealogical research and resources for research. Pierre Durye, an eminent genealogist and former chief custodian of the Archives nationales, is the author of many works on the history and the genealogy. *Genealogy: An Introduction to Continental Concepts*, trans. by Wilson Ober Clough is considered to be the best book in English on French genealogy. Excellent section about search strategies for finding notaries and notorial records. Bibliography: p. 135-136. Includes index.
7782714

France: research outline. Salt Lake City: Family History Library, 2002. 47 p. : map.
CALL NUMBER: CS583 .F73 2002
      Online at: http://www.familysearch.org/Eng/Search/Rg/guide/France.asp#introduction
Discusses records for French genealogical research, and has bibliographies of books about each kind of record.
1996 edition: 97163360
2002 edition: 2004273636

Konrad, J. ***French and French-Canadian family research***. Munroe Falls, Ohio: Summit Publications, 1985. ii, 76p., ill.
CALL NUMBER: CS83 .K65 1985
      The chapter, "Ancestor Hunting in France," explains how to use sources to determine the name of the town or province from which the French ancestor immigrated. Useful for North Americans are short sections about French who fought in the American Revolution, and the immigration from Alsace-Lorraine. Descriptions of civil registration (vital records), Catholic parish records, Protestant and Huguenot records, censuses, notarial records, and French army records. Lists of names and addresses of departmental archives and French genealogical societies (1985). Examples of sample letters in French which request information from France. A useful French vocabulary list.
88178238

Lichliter, Asselia S. **"Research in France."** <u>National Genealogical Society Quarterly</u>. 74: March 1986, p. 49-57.
CALL NUMBER: CS42 .N4<LH&G>

Discusses records to be found in France in archives and libraries in Paris, and in the 90 departments. Discusses reasons for migration to the New World, importance of research in records at home, military records, the Bibliothèque Nationale, the Archives Nationales, notarial records, the Archives d'Outre Mer (Overseas Archives containing records of individuals who migrated from France to places overseas), the Departmental Archives of Paris, Independent depositories in Paris such as the Historical Society of French Protestantism (La Société du Protestantisme français) the Bibliothèque Mazarine, the Archives of the Israelite Consistory of France and Algiers, town halls in Paris, department offices outside Paris.

Websites:

Cyndislist: France
http://www.cyndislist.com/france.htm
A list of web sites about French genealogy.

French Genealogy & Family History
http://genealogy.about.com/od/france/French_Genealogy_Family_History.htm
"Search for your French and French-Canadian ancestors in this collection of genealogy and family history databases and resources for France. Includes tutorials for researching French ancestors, suggestions for writing to France and translating French records, and information on civil records, parish registers and other French genealogical records."

FamilySearch Research Wiki
France
https://wiki.familysearch.org/en/France
FamilySearch Wiki is a community website dedicated to helping people throughout the world learn how to find their ancestors. Through the France Portal page you can learn how to find, use, and analyze French records of genealogical value. The content is variously targeted to beginners, intermediate, and expert researchers. Here you will find helpful research tools and research guidance.

Genealogy in France: A Beginner's Guide to Researching Your French Ancestry
By Kimberly Powell, About.com Guide
http://genealogy.about.com/od/france/a/french_ancestry.htm
Includes tutorials for researching French ancestors, suggestions for writing to France and translating French records, and information on civil records, parish registers and other French genealogical records.

2.   French

Archassal, Pierre-Valéry. *Généalogie sur Internet*. Paris: Campus Press, 2001. vii, 289 p., ill.
CALL NUMBER: CS14 .A68 2001
2002423499

Archassal, Pierre-Valéry. *Généalogie, une passion moderne*. Paris: Bourin, 2006. 236 p.
CALL NUMBER: CS10 .A73 2006
2006422730

Archassal, Pierre-Valéry. *Mémento de paléographie généalogique*. Paris: Brocéliande, 2000. 63 p.
CALL NUMBER: Z115.F72 A73 2000
   A French genealogical handbook about the topic of French paleography, or hand-writing.
2001366104

Aublet, Robert. *Nouveau guide de généalogie*. Ouest-France: 1986. 189 p., ill., map.
CALL NUMBER: CS10 .A93 1986<LH&G>

   Discusses descendant and ascendant genealogy, including charts and numbering systems for organizing genealogical information. Includes research strategies for *état civil*, decennial tables, parish registers, Protestant and Jewish records, notarial documents, tax and military records. Several case studies give examples of the use of these documents. Discusses National Archives. Also communal, departmental, ecclesiastical and notarial archives, as well as useful libraries. Also, calendar conversion tables, examples of antiquated handwriting, genealogical terminology, conventional genealogical abbreviations.
87171979

Beaubestre, Jean. *Précis de généalogie*. Paris: Christian; Toulouse: Editions bv Publications, 1996. 227 p. bibliographical references.
CALL NUMBER: CS10 .B39 1996
96214941

Beaucarnot, Jean Louis. *Chasseur d'ancêtres*. Paris: Menges, 1980. 293 p.: ill.; bibliography.
CALL NUMBER: CS10 B4
   Discusses types of genealogy: classical genealogy of nobility and armigeous persons; professionals who trace genealogy in order to identify those who may have inherited property; and modern genealogy, which is more democratic. Included are sections on the meaning of family names, and on ascendant and descendant genealogy. Section about how to search discusses the civil and parish vital records. Devises a search schema for tracing genealogy, with elements A.B.C.D.: the act (vital records such as birth, marriage or death, and parish records such as baptism, marriage and burial); the "bonhomme" (or the identity of the ancestor concerned); the commune (where the act is registered); and the date (at least the approximate year). Practical sections summarize the contents of birth, marriage and death certificates, describe annual and decennial tables, show how to systematically record information, and explain degrees of relationship under civil and canon law.
80124337

Beaucarnot, Jean-Louis. *Trésors et secrets de la généalogie : mémoire patrimoine, noms de famille, ancêtres, racines, archives, souvenirs, dynasties*. Paris: JC Lattès, 1998. 223 p.: ill. (some col.).
CALL NUMBER: CS17 .B415 1998
   Bibliographical references, p. 22
99172447

Beaucarnot, Jean-Louis. *Généalogie mode d'emploi*. [Alleur, Belgium?] : Marabout, 2002. 128 p.: ill. (some col.).
CALL NUMBER: CS17 .G414 2002
2002476933

Beaucarnot, Jean-Louis. *Qui étaient nos ancêtres? : de leur histoire à la nôtre*. Paris: Latte`s, 2002. 426 p.
CALL NUMBER: CS583 .B43 2002
 2003418679

Callery, Pierre. *La généalogie, une science, un jeu: quelques éléments techniques pour une recherche généalogique*. Paris: Seuil, 1979.
CALL NUMBER: CS17 .C34
    A short explanation of genealogical research in France, ranging from family oral history and cemetery visits to vital records and the records of notaries. Discussion of the arrangement of genealogical research, including numbering and charts, under the title "Classification of Discoveries." Blank charts to arrange ancestors back to the 15th generation.
81464195

Colin, Michel. *Généalogie ascendante et descendante informatisée: méthode Gadi*. Caen: M. Colin, 1985. 225 p., ill.
CALL NUMBER: CS583 .C447 1985
    An historical analysis of French record-keeping, including vital records (Catholic parish, Protestant, Jewish and civil), annual and decennial indexes to civil records; cemetery records, electoral lists, military records, naturalization and emigration records, and how to use these records for genealogical research. Discusses French institutions important in record keeping, such as the court and taxation system, notaries, and other French office holders.
    Description of institutions with genealogical records: communal archives, departmental archives, notarial archives and the National Archives. Also describes libraries where genealogical information can be obtained: the Bibliothèque Nationale, municipal libraries and special libraries. Includes a Latin-French word list, and examples of early handwriting to help in deciphering early records.
86155899

Dubourguey, Christian. *La généalogie: guide pratique pour la recherche de vos ancêtres et de votre famille.* [Monte Carlo]: RMC edition, 1988.
CALL NUMBER: CS583 .D84 1988
    Discusses theory of ascendant and descendant genealogy. Discusses the origin of geographic names and surnames, and of personal names. Compares genealogy and demographic history. Describes records in communal and departmental archives, the National Archives and libraries. Discusses specialized archives, such as notarial archives, ecclesiastical archives, military archives and ecclesiastical archives. Describes vital records, civil as well as Catholic parish, Protestant, and Jewish. Practical sections about genealogical signs and symbols, abbreviations, organization of information, genealogical numbering systems, and tables. An appendix discusses geographical arrangement of about 100 common names by region.
88-130625

Du Passage, Yves. *Mes aïeux, quelle histoire: guide pratique de la généalogie pour tous*. [Paris]: Hachette, 1986. 191p., ill.
CALL NUMBER: CS583 .D8 1986
    Discusses compiling a genealogy. Lists kinds of genealogists, different forms of genealogies and numbering systems. Discusses use of private sources such s the "livret de famille," (a family book kept in families listing about three generations, photographs, and oral history). Emphasizes the use of public

records to verify private sources. Discusses contents of the various archives. A chapter is devoted to public sources in the Ancien Régime, especially the vital records: Catholic parish records, Protestant and Jewish records. Gives a practical search strategy for finding notarial records if know only two of three necessary elements for the search: name of notary, place of residence, and date of act being researched (p. 126-128). Practical examples of search strategies, illustrated by case studies. Includes calendar conversion tables, and a short discussion about heraldry.
86-221356

Grandeau, Yann. *A la recherche de vos ancêtres: guide du généalogiste amateur*. Paris: Stock, 1974.
CALL NUMBER: CS583 .G72
    Discusses genealogy as a science. Includes preliminary investigations such as oral traditions, family papers and published or printed sources. Discusses vital records (l'état civil), parish registers, Protestant and Jewish registers, and notarial registers. Emphasizes not only the content of documents, but how to deal with difficulties and exceptions. There is a bibliography at the end of each chapter, as well as a compiled bibliography at the end of the book.
75521084

Henry, Gilles. *Recherchez vos ancêtres: guide de recherches généalogiques*. Conde-sur-Noireau: Editions C. Corlet, 1982. 129 p., ill.
CALL NUMBER: CS584 .H46 1982
    Discusses ascendant and descendant genealogy; civil registration, and notarial documents. There are many transcribed documents. Lists the French departments, and the corresponding ancient provinces from which they were formed. Lists departmental archives (with addresses and hours) and genealogical societies. In the Preface, lists the eight most useful books published on the subject of genealogy (prior to 1981).
83117811

Henry, Gilles. *Retrouver ses ancêtres, c'est facile*. Avec la collaboration de Emmanuelle Héaume. Paris: A. Michel, 1997. 235 p., ill.
CALL NUMBER: CS10 .H28 1997
99519458

Hezelles, Norbert. *La Généalogie*. Paris: Hachette, 1979. 239 p., ill., bibliography.
CALL NUMBER: CS583 .H49
    Presents a practical method to find ancestors, including keeping precise notes and organization of information. Discusses vital records (parish, Protestant, Jewish) and the elements of information that are contained in birth, marriage and death records. Emphasizes problem solving to deal with old handwriting, lacunae in registers and errors in names, places and relationships. Presents search strategies. Discusses ascendant and descendant genealogies, and presents tables for arranging information. Gives useful suggestions for compiling a family history, including compiling studies of historical and geographical origins of the family, and the application of statistical methods to the study of genealogy. A short lexicon of definitions of French genealogical terms. Bibliography, p. 238-239.
79-377706

La Généalogie: histoire et pratique. Edited by Joseph Valynseele. Paris: Larousse, 1991. 325 p., ill.
CALL NUMBER: CS10 .G46 1991
    This is an encyclopedic study of the multiple facets of genealogy: its long history, reasons why it has become popular with contemporaries, sources, methods. Edited by LaRousse, it is the result of

collaboration of 25 authors, all specialists in, or practitioners of, genealogy. It covers the philosophic: the relationship of genealogy to history, heraldry, biography, literature, the arts, sciences, pedagogy, demography and medicine. It covers the practical, with descriptions of vital records, the Archives Nationales, the departmental archives, and the various libraries. There are sections about Protestant, Jewish, Alsatian, Belgian, Swiss and French Canadian research. There is a bibliography at the end of each chapter. Includes a list of sources and addresses.
92131901

Jouniaux, Leo. *Généalogie: pratique, méthode, recherche*. Paris: Arthaud, 1991. 415p., ill.
CALL NUMBER: CS10 .J68 1991<LH&G>
  What sets this handbook apart is the theory, the detail of the examples, the excellent charts summarizing material covered in chapters, and the consistent suggestion of logical search strategies for many genealogical problems. Theoretical sections about use of oral tradition, critical review of documents, testing of hypotheses, schema for the organization of genealogy, including genealogical numbering systems. Discusses relationship of genealogy to other social sciences. An important summary of the institutions of France, both administrative and ecclesiastical to help the non-French researcher understand terms in documents, as well as the provenance of documents encountered in research. Practical sections include: glossary of French genealogical terms, examples of handwriting and abbreviations, information about the transformation of geography from provincial to departmental, calendar conversion, units of measure (including old money), and maps. Discusses laws governing archives and archival deposits, descriptions of communal (municipal) archives, and departmental archives. Describes libraries and museums that include information useful to genealogists. Also, describes archives in Paris, including the Archives Nationales, archives hospitalières, and the Archives of Paris.
  Records covered include: ètat civil, Catholic parish, Protestant, and Jewish records, notarial records, and judicial records.
91-215155

Mergnac, M. O. (Marie-Odile). *Ma généalogie de siècle en siècle*. Paris: Archives & culture, 2009. 271 p. : ill. (some col).
CALL NUMBER: CS17 .M47 2009
  "Enfin un guide de généalogie qui n'est pas construit en fonction des documents à consulter mais des questions qu'on se pose ! Ce livre part de vos envies, de vos questionnements, de vos doutes, de votre période de recherche, aussi loin soit-elle dans le temps, pour vous faire progresser et franchir les écueils, siècle par siècle. Accessibles à tous, les fonds d'archives en France sont d'une telle richesse qu'il est possible non seulement de retrouver les noms de vos ancêtres jusqu'au XVIe siècle, mais aussi « d'entrer » virtuellement dans la maison que l'aïeul occupait il y a trois cents ans, de retrouver ses brouilles avec ses voisins ou ses aléas de fortune, comme si vous l'aviez connu. Lancez-vous donc dans l'aventure généalogique : elle ne demande que du temps et peut se pratiquer avec ou sans budget, sur place ou à distance, par tous les temps et quel que soit votre âge : un passe-temps intellectuel passionnant qui rassemble toutes les générations!" Review by publisher.
2009492060

Pouye, Bernard. *La généalogie, comment, pourquoi?* Paris: Centurion, 1982. 78 p., ill.
CALL NUMBER: CS583 .P68 1982
  Short descriptions of sources: vital records (etat civil), parish registers, and the acts of notaries. Describes the Archives de l'Enregistrement, where contracts were registered and taxed; the Archives Judiciaires, which hold traces of legal proceedings in which our ancestors were embroiled; the Archives

Militaires, which holds military records other than the military service records (the latter are held in the departmental archives), Archives de l'Assistance Publique (records about public assistance, whether about social assistance, or sanitation); Archives d'Outre-Mer (vital records about French born abroad, but also about the French administration of the colonies); Archives religieuses (parish records, records of bishops, abbeys, and ecclesiastical courts). Describes methods of arranging genealogies in ascendant and descendant order, including genealogical numbering systems.
A short theoretical chapter about why people do genealogy, and how it relates to other fields (history, chronology, genetics and philosophy).

Appendix gives extracts of records, conversion table for Republican and Gregorian calendars, and a short annotated bibliography.
83121624

Thiébaud, Jean-Marie. ***Pratique de la généalogie: guide universel de recherche***. Besançon: Cêtre, [1995]. 322 p., ill., bibliographical references.
CALL NUMBER: CS17 .T48 1995
95159480

B. Archives

France has three major types of genealogical record repositories: The National Archives; departmental archives; and town registrars. There are also libraries, including the Bibliothèque Nationale, and other smaller libraries and archives. The following books describe archival records, in general, and the archives and libraries in which they are deposited. In addition to the more general book about genealogical research for all French families, there are two books about the genealogical records for tracing members of the two largest non-Catholic religious groups: Protestants and Jews. There is a substantial literature about noble families under the subject of Heraldry (not included here). However, there are some references for books about tracing noble families included here.

Bernard, Gildas. ***Guide des recherches sur l'histoire des familles***. Paris: Archives nationales, 1981. 335 p.
CALL NUMBER: Z 5313 .F8 B47 1981<LH&G>

This guide, by the then Inspector General of the Archives of France, establishes a survey of all the different categories of documents for genealogical and biographical research in France, indicates where the documents can be found, and explains how one can obtain access to them. It surveys both public and private archives, from the Archives Nationales to departmental archives, as well as specialized archives such as the Archives of the Army and those of Foreign Affairs. It also includes resources at the Bibliothèque Nationale.

Discusses the organization of the public archives in France; the état civil; diverse sources that complement vital records, such as naturalizations, legitimations, name changes, etc.; notarial documents and the various types of archives where they are deposited; population censuses and electoral lists; military archives; fiscal archives (taxes); judicial and police archives, and cultural archives. From the above categories of documents which concern all families in France, Bernard passes to those which concern special categories: functionaries, and especially under the Ancien Régime, officers; diverse careers which the Archives documents well, such as artists, certain categories of intellectuals, doctors, pharmacists, and ministerial officers; the nobility, and decorated individuals; heraldry.
82177953 r92

Colin, Michel. ***Les archives des Français***. Bois Guillaume: M. Colin. Condé-sur-Noireau: Diffusion, Editions Charles Corlet, 1998. 350 p., ill.
CALL NUMBER: CS591 .C65 1998 France<European RR Reference>
99190593

Diocesan Archives
http://www.afg-2000.org/archives_diocesaines/sommaire.html
    Click into geographic area on map to get a list of archives and their addresses. Also gives particular information about certain dioceses, (how each is geographically derived from one another and which communes are contained ,etc.).

Gadille, Jacques. ***Guide des archives diocésaines***. Lyon: Centre d'histoire du catholicisme, 1971. 167 p., maps.
CALL NUMBER: CD 1218 .A2G3
    Gadille presents a preliminary inventory of the most important Catholic church records and registers in diocesan archives in France. This preliminary inventory is not meant to substitute for more detailed inventories, but it is useful for its summarization of the types of documents contained in each diocesan archive. Since the inventory was sponsored by the Secretary General of the Bishopric (Secrétariat général de l'episcopat), the Society for the ecclesiastical history of France, and the Center for the History of Catholicism at Lyon, it was possible to obtain information from almost all the diocesan archives (although not all archives provided the same detail in their responses). The book is arranged alphabetically by diocese. Gadille describes which indexes or inventories exist for each diocesan library. He also describes the format of the indexes (published, card files or otherwise at the diocesan archives). He differentiates which material has been deposited in the departmental archives. In addition, one of the appendixes is an index to personal papers deposited in diocesan archives.
    The documents are arranged in systematic categories under each diocesan archive (although not all archives have the same documents). The most interesting documents for genealogists tend to be under the heading "Sociologie du clergé et du peuple chrétien." These documents include, among others, the following: (1) Parish registers (Registres de catholicité des paroisses); (2) Alphabetical lists of priests and clerical orders; (3) lists of priests during the French Revolution--those persecuted, or who were victims during the Revolution, those who remained faithful to the church, those that emigrated to Switzerland, etc. (4) for some dioceses, dossiers exist for each priest; these are sometimes enriched by personal papers of many of the priests; (5) sometimes there are dossiers about seminarians and laics as well; (6) registers of ordinations (these are generally conserved in series G of the archives départementales, and certain ones are published, but the archives diocésaines conserve some old series--listed on page 15-16); (7) resignations of clergy; (8) dispensations from religious duties (for fasting, etc.); (9) questionnaires before confirmation; (10) obituaries of clerics; (11) parish newsletters and newspapers (for example, from the mid-19th century when <u>Semaines Religieuses</u> were generalized, these official newsletters of dioceses carried necrologies). (12) documents relating to clerical, canonical or pastoral visits to parishes (an appendix gives a table of diocesan visits, reports and inquests by region and time period); (13) inquests into religiously mixed marriages; (14) compiled histories of parishes (some are manuscripts dating to the 18th century).
73326991

Jensen, C. Russell. ***Preliminary survey of the French collection***. Salt Lake City: University of Utah Press, 1980. xxxix, 433p.
CALL NUMBER: Z5313 .F8J46

This is a guide to French records on microfilm housed at the Family History Library in Salt Lake City. This preliminary survey (1980) of French records microfilmed by the Church of Jesus Christ of Latter-day Saints gives precise information about vital records (parish and civil records) held at the LDS family History Library in Salt Lake City, Utah. Records are listed geographically, by department, and then by town. Indicates regions where records are missing, as well as those where there are extensive holdings of microfilmed records. Other types of records microfilmed, that supplement vital records for genealogical information, such as census, emigration and immigration, and notarial records, are listed in Appendix A.

As microfilming has continued past 1980, this preliminary survey should be supplemented by use of the Family History Library Catalog, which is available at http://www.familysearch.org

It is possible to order microfilmed records through local Family History Centers. See: http://www.familysearch.org
8020810 r903

Wolff, C. *Guide du généalogiste aux archives departementales et communales*. Versailles: I.E.E.G., 1983. 67 p.
CALL NUMBER: Z5313. F8 W64 1983

Although the descriptions of services of departmental and municipal archives in France (such as addresses, hours, research by correspondence and reproduction of documents) may have changed, this is a useful book to use as: (1) a list of departmental and municipal libraries. and (2) it describes the availability of vital records (both parish and civic records) as well as formats available (print, microform or original) and indexes (such as decennial tables) available for departmental, municipal and communal archives.

Many of these archives now have web sites
86-122990

Departmental Archives in France: Websites:
   UNESCO Archives Portal: An international gateway to information for archivists and archives.
http://www.unesco-ci.org/cgi-bin/portals/archives/page.cgi?g=Archives%2FGovernment%2FState_and_Regional%2FEurope%2FFrance%2Findex.html;d=1

1.   **Notary Records**

Ardouin-Weiss, Idelette. *Les actes notariés anciens: lexique*. Tours: Centre généalogique de Touraine, 1991. 58 leaves, ill.
CALL NUMBER: KJV187 .A85 1991 FT MEADE

Legal language of notaries in French history. Series: Collection Centre généalogique de Touraine. Includes bibliographical references (p. 57).
94148364

Montjouvent, Philippe de. *Dépouiller les archives de notaires*. Paris: Autremont, 2004. 80 p.
ON ORDER
   Handbook about how to use notorial records. The notorial records are a multiple source of

information for all French families. Records notarized include such as marriage contracts, wills, inventories, property records, and sales and gifts.

French Genealogy Blog
http://french-genealogy.typepad.com/genealogie/notaires/
    Includes: "What Is A Notaire?" "Notorial Records," and "Finding Notorial Records"
    Anne Morddel, 18 February 2010.

2. Protestant Families

Arnaud, E. (Eugène). *Émigrés protestants dauphinois, secourus par la Bourse française de Genève de 1680 à 1710.* Grenoble: Impr. et lithographie F. Allier, 1885. 66 p.
CALL NUMBER: BX9456.D3 A76
    List of names extracted from a ms. entitled: Liste des assistés de la Bourse française de 1680 à 1710, by J. C. Auquier. Supplements the author's Histoire des protestants du Dauphiné.
74173506

Bernard, Gildas. *Les familles protestants en France: XVIe siècle-1815: guide des recherches biographiques et généalogiques.* Paris: Archives nationales, 1987.
CALL NUMBER: Z5313. F8 B46 1987
    There is a short history of Protestantism in France which helps illuminate the problems in finding Protestant records in the Ancien Régime prior to the establishment of the civil registration in 1792. Protestant pastors kept vital records from 1559 to 1685. After the revocation of the Edict of Nantes (1685) until the Edict of Toleration (1787), there are a certain number of clandestine records, "au Désert," as they are called. From around 1740, itinerant pastors traveled around certain regions, which often means that the records of more than one departmental archives must be searched.
    Besides Protestant pastors' registers of vital records, supplementary records discussed include: burial registers, individual and group abjurations (denial of the Protestant religion), lists of new converts to Catholicism and their children, registers of Catholicity (in which Protestants who renounced Catholicism on their death bed, or upon marriage, are listed), lists of fugitives or of prisoners and their property, and records of Protestant notaries who notarized the major part of the acts of their co-religionists.
    Documents useful for French Protestant genealogy are arranged by specific archive or library (important since many are not in the local departmental archives where one would think to search, and some are even abroad in places like Switzerland). Also includes a bibliography of printed works about the history of Protestantism in each department, and detailed archival guides.
88192129 r92

Christian, Francis. *Retrouver ses ancêtres protestants.* Paris: Editions Autrement, 2005. 64 p., ill., map.
CALL NUMBER: CS596.P75 C47 2005
2005431440

3. Jewish families

Bernard, Gildas. *Les familles juives en France, XVIe siècle-1815: Guide des recherches biographiques*

et généalogiques. Paris: Archives nationales, 1990. 281 p., bibliography.
CALL NUMBER: Z6373 .F7 B47 1990

    Archival resources about French Jewish families listed by department. Before the civil registration of vital records by the state (l'état civil), there were few vital records for French Jews. After civil registration was adopted in 1792, and after all French had to adopt surnames in 1808, vital records exist for Jews along with all other French people. This book discusses the earlier vital records that do exist, as well as substitutes for these records that can be found in archives: registers of burials of "non-Catholics" (1685-1792), registers of the Edict of Tolerance (1787), letters of naturalization, tombstones, notarial archives, conversions to Catholicism, and registers adopting a surname. Essays about the history and demography of Jews in France, as well as about the geographic areas of France with the highest Jewish population: Alsace, Lorraine, Avignon and Venaisson in the Department of Vaucluse, as well as in the Southwest of France.
91124438

Abensur-Hazan, Laurence. ***Rechercher ses ancêtres juifs***. Paris: Autrement, 2005. 80 p., ill.
CALL NUMBER: CS596.J4 A34 2005
2006403452

Ginger, Basile. ***Guide pratique de généalogie juive en France et à l'étranger***. Paris: Cercle de généalogie juive, 2002. 278 p.: maps.
CALL NUMBER: CS596.J4 G56 2002
2003432204

Katz, Pierre.
Les communautés juives du Haut-Rhin en 1851: relevés du recensement. Paris: Cercle de généalogie juive, 2002. 1 v. (unpaged) : 1 map.
CALL NUMBER: DS135.F85 H395 2002
2004392981

4. Nobility

Bibliography for Tracing French Noble Families
by John P. DuLong
http://habitant.org/tools/noblebib.htm

French Heraldry and Related Topics
    Table of contents of the pages devoted to French heraldry, nobility, royalty, knighthood.
http://www.heraldica.org/topics/france/

Montjouvent, Philippe de. ***Retrouver ses ancêtres nobles***. Paris: Autrement, 2005. 96 p. : ill.
CALL NUMBER: CS587 .M66 2005
2005476436

Nobility and Titles in France
http://www.heraldica.org/topics/france/noblesse.htm

5. **Bibliographies**

The following national bibliographies of French genealogy, nobility, heraldry and local history are multi-volumed and fairly exhaustive up to the date of publication. The researcher may use these national bibliographies by Arnaud and Saffroy to find published genealogies of French families that appeared as books or journal articles before 1982.

Arnaud, Etienne. ***Répertoire de généalogies françaises imprimées***. Paris: Berger-Levrault, 1978-1982. 3 vols.
CALL NUMBER: Z5305 .F7A75
Bibliography of French genealogies printed before 1982 which the author states is "as exhaustive as possible, but without the least illusion that lacunae do not exist." It includes genealogies and fragments which include at least three consecutive generations (although if a shorter fragment is the only one on a family name, he includes it). This is a systematic search for the genealogies of French families in France, but it also includes some genealogies from French speaking countries, especially Canada, Belgium and Switzerland. All categories of families are included: nobility, bourgeois, working class; Catholic, Jewish and Protestant. Volume 2 includes a supplement to volume 1; and volume 3 includes a supplement to volumes 1 and 3. This bibliography has supplements, but does not attempt to duplicate Saffroy.
78385399 r862

Gandilhon, René. ***Bibliographie générale des travaux historiques et archéologiques publiés par les sociétés savantes de la France, dressée par René Gandilhon sous la direction de Charles Samaran. Période 1910-1940.*** Paris: Impr. nationale, 1944-61. 5 v.
CALL NUMBER: Z2183 .G3
Contents: t. 1. Ain-Creuse.--t. 2. Dordogne-Lozére.--t. 3. Maine-et-Loire-Haute-Savoie.--t. 4. Seine.--t. 5. Seine-et-Marne-Yonne. France d'outre-mer et étranger. Notes: Continues Lasteyrie's Bibliographie générale des travaux historiques et archéologiques publiés par les sociétés savantes de la France (including its supplement, Bibliographie annuelle, which covers the literature published through 1910).
47021983

Poull, Georges. ***Les cahiers d'histoire, de biographie et de généalogie; travaux historiques, études généalogiques et documents inédits ou méconnus, Rupt-sur-Mosell (Vosges)***, 1965-(72). v. (1-6).
CALL NUMBER: DC5 .P6
Six volumes of collected works (bibliography) about French local history and genealogy.
66057496

Saffroy, Gaston. ***Bibliographie généalogique, héraldique et nobiliaire de la France des origines à nos jours, imprimés et manuscrits***. Paris: G. Saffroy, 1968-88.
CALL NUMBER: Z5305 .F7S22
A national bibliography of French Genealogy, heraldry, nobility and local history. Contains books, manuscripts and periodical articles. Volume I lists general books about French genealogy, French history, and French institutions. Volume II is arranged by province, and lists books relating to the history of the province, its families and its institutions. Volume III includes collective genealogies as well as genealogies of particular families. It covers noble as well as non-noble families. Volume IV is a bibliography of sources, arranged by author, title and subject. Volume V is a supplement covering 1969-

1983.
68136872

C. Biographical Dictionaries

France has a long tradition of biographical publishing. This section lists bibliographies of French biographies, and a few universal French biographical dictionaries, and a representative sample of the large number of biographical dictionaries that treat French figures held by the Library of Congress. There are many specialized biographical dictionaries that treat a certain social class, a certain profession, or type of office holder. Because the Library of Congress collection is historic, some biographical dictionaries concern themselves with figures who lived long before the 20th century, while other dictionaries only concern 20th century figures. Biographical dictionaries not listed here may be found under the various Library of Congress subject headings including those for geographic areas of France and professions.

Bradley, Susan ed.
Archives biographiques françaises.   London: Bowker-Saur Ltd., 1988 (1990?).--1073 microfiches with paper index.
CALL NUMBER: Microfiche 90/7007
    A compilation of biographical notices, selected from 180 biographical works, about circa 140,000 people in France, French speaking Switzerland, Belgium, French Canada, and the Francophone colonies, as well as foreigners associated with France for a period of their lives. Not only the rich and famous have been included. Sailors, mechanics, engravers, printers and binders, industrialists, agriculturalists, missionaries, policemen and many others from all walks of life are represented. The span of coverage is from the beginning of French civilization to 1914.
    ""Fusion dans un ordre alphabétique unique de 180 des plus importants ouvrages de référence biographiques français publiés du 17e au 20e siècle"--Jacket. Accompanied by: Liste des sources."
90956049

CALL NUMBER: Paper index, Guide to Microfiche 90/7007: CT1003.I53 1993, *Index Biographique Française*. edited by Helen and Barry Dwyer. London: New Jersey: K. G. Saur, 1993. 4 vols. Request in Reference, Main Reading Room, Microfilm Guides.
    The *Index Biographique Française* provides a summary of the information contained on about 140,000 people featured in the *Archives Biographiques Françaises*. This index serves a dual purpose: it is an index to *Archives Biographiques Françaises*: the user can tell whether or not a particular individual appears in the Archives and if so where the information is to be found and how detailed it is likely to be. It is also an independent research tool: the basic biographical data provided will almost always suffice for the purposes of identification, and if further information is required, the user is directed to the original source from which the entry has been compiled.
    Provides 40,000 cross references for individuals known by more than one name. It also provides a list of sources indexed, which can serve as a bibliography of important French biographical sources. These sources are accessible in the Microfiche 90/7007 collection, *Archives biographiques françaises*. Many of them are also in book form in the General Collection at the Library of Congress.
 94119649

Archives biographiques françaises. Deuxième série. Nappo, Tommaso, ed. Munchen: K.G.Saur Verlag, (1993?).--644 microfiches.
CALL NUMBER: Microfiche 97/13

Brings together into a single alphabetical arrangement entries from 122 of the most important biographical reference works for persons living during the 19th and 20th centuries through the 1960's.
96195609

Index biographique français. (*Französischer biographischer Index*). (French biographical index). Nappo, Tommaso, comp. 2ème éd. cumulée et augm. München : K. G. Saur, 1998. 7 v. ; 30 cm.
CALL NUMBER: Microfiche 90/7007

Serves as an index to the microfiche set entitled: *Archives biographiques françaises*. Includes bibliographical references (p. xvii-xxiii). French, with preface and notes for use in English, French, and German. Contents: 1. A-Brasquet -- 2. Brass-Decye -- 3. Dedaux-Gaujot -- 4. Gaujour-La Helle -- 5. Lahens-Mieg -- 6. Miège-Ripart -- 7. Ripault-Z
98111890

Dictionnaire de biographie française. Balteau, J., Barroux, M., Prévost, M., eds. Paris: Librairie Letouzey et Ané, 1933-[2009]. v. <1-13, 112-118 >.
CALL NUMBER: CT143 .D5 Biog [MRR]

Issued in fascicles. Vols. 4-8 "sous la direction de M. Prévost et Roman d'Amat."; v. <9-13 > "sous la direction de Roman d'Amat."; v. 112-113 sous la direction de M. Prévost ... [et al.] Includes bibliographical references. LC holdings: Unbound fascicles are recorded until bound, and are located in the Cataloger's Reference Collection (Copy 1) and the Main Reading Room (Copy 2). Volume XIX, Fasc. CXIV in process, as of Sept. 19, 2001.

Alphabetical arrangement. Included are selected persons from ancient Gaul forward. Excluded are living persons who were alive as of about 1925. This dictionary does not include the colonies or the French African departments, Corsica, or Alsace. But it does include French Flanders, the Franche-Comté, Lorraine, Provence, Savoie or Comtat Venaissin, even though these places have not always been a part of France. People who are an important part of French history (although not born in France) are included.
33010965

Fierro, Alfred. ***Bibliographie analytique des biographies collectives imprimées de la France contemporaine: 1789-1985.*** Préface de Michel Fleury. Paris: Libr. H. Champion, 1986. vii, 376 p.
CALL NUMBER: AS162.B6 Fasc. 330 France. [European Reading Room]

An analytical bibliography of 2700 collective biographies published 1789-1985, that contain biographical notices primarily about late 18th century to 20th century French people. The work is divided into three parts: (1) general biographical works, subdivided chronologically; (2) biographical works which concentrate on professions; and (3) biographical works that concentrate on regional or local areas. If a work fits under more than one category, there are cross references to the main entry. The entries are annotated.

Fierro, a paleographic archivist and conservateur at the Bibliothèque Nationale, compiled this bibliography by systematically gathering references from the principal bibliographies (He lists Besterman, Lobies, Slocum as well as five bibliographies of biographical works that are narrower in scope than this national one), from printed catalogs of the largest libraries in France, and the works held at the Bibliothèque Nationale. Included are biographies of the French, but not where less than 25% of the material is about French people. Covered is the period 1789-1985, but excluded is the period before 1789. Formats included are alphabetical dictionaries (but not encyclopedias), and excluded are biographical notices that are dispersed throughout the text of a book, and only listed in the index of that book. Included are those with contents with a minimum of three biographies that have dates of birth, date and profession, while excluded are individual biographies or those limited to two persons, partial biographies limited to an episode or two, and notices limited to name, address and profession, such as most of the professional directories. Included are printed books with a diffusion of at least 100 copies and excluded are manuscripts, typewritten or mimeographed works with less than 100 copies, as well as documents on magnetic tape. This work does not systematically screen or index periodicals, but does list periodicals where the dominant theme is biographical.

Fierro compares this bibliography to other important French bibliographies of biographical works: (1) The only important French work which includes biographies of contemporaries is le *Dictionnaire de biographie française*, which began publication in 1933, but never got beyond the letter G, and did not include notices about all the personnages that played a role in French history since 1789. (2) Two books of exceptional value (listed under Bibliographies in this publication) which can be used for biographical research, but which are primarily focused on genealogical research are la *Bibliographie généalogique héraldique et nobilière de la France* by Gaston Saffroy (Paris: 1968-1979), and le *Répertoire de généalogies françaises imprimées* by Etienne Arnaud (Paris 1978-1982). (3) The 500 biographical works that are solely about France, and cover the last two thousand years of French life, listed among the 5,000 works about world biography in l'Index bio-bibliographicus by Jean-Pierre and François Pierre Lobies.

This work will guide readers to the important biographical dictionaries from 1789-1985. Many of these works are in the Library of Congress collections.
87149379

Hoefer, Jean Chrétien Ferdinand. ***Nouvelle biographie générale depuis les temps les plus reculés jusqu'à nos jours, avec les renseignements bibliographiques et l'indication des sources à consulter.***
Paris: Firmin Didot frères, fils et cie,, 1853-66. [v. 1, 1857]. 46 v.
CALL NUMBER: CT143 .H5
   Biographical Dictionary; Bio-bibliography.
02003498

Lobies, Jean-Pierre, François-Pierre Lobies, Otto Zeller and Wolfgang Zeller, eds.
IBN: Index bio-bibliographicus notorum hominum. Osnabruck: Biblio Verlag, 1973-2001.
CALL NUMBER: Z5301 .L7 (Main Reading Room)
   Index bio-bibliographicus lists over 5000 biographical works on a worldwide scale, of which there are 500 solely about France. These biographical works cover the last two thousand years of French life. Part B lists the biographical references arranged by geographical, historical and linguistic principles. At the end of Part B is an index, which directs the reader to French sources under the general term France. In Part B, there are French references scattered throughout the text because while the first arrangement is geographical (and there is a section on France), there are other French biographical dictionaries under other categories (such as each profession).
Description:
   Volumes: B (2 v.); C, I, 1-98, 99.1-2, 100-111, 114; III, 1-4; VI, 2; in 118 > ; 26 cm.
Incomplete Contents:
   Pars B. Liste der ausgewerteten bio-bibliographischen Werke -- Supplementum -- pars C. Corpus alphabeticum. I. Sectio generalis <v. 1-98, 99.1-2, 100-111, 114 > III. Sectio Armeniaca <v. 1-4 > VI. Sectio Sinica cum supplemento Coreano. <v. 2 >
Notes: Edited 1990-1997 by Otto and Wolfram Zeller. Edited 1997-<2001 > by Wolfram Zeller.
   Vol. 99.1 has spine title: Pars B. Catalogus operum examinatorum, supplementum no. 6216-6242. Pars C. Corpus alphabeticum, sectio generalis, vol. 99.1. Supplementum A.
Vol. 99.2 has spine title: Pars B. Catalogus operum examinatorum, supplementum no. 6243-6247. Pars C. Corpus alphabeticum, sectio generalis, vol. 99.2. Supplementum B-Bi.

Vol. <99.2, 107-111, 114 > published by Felix Dietrich Verlag.
   The Library of Congress does not have Part A, which includes a list of evaluated bio-bibliographical works, dealt with in a more detailed way.
   Part C is an alphabetical name index which guides the reader to biographical sources listed in Part B. However, it is not finished (the last volume received by 1988 is Vol. 88, which ends in names starting with GIO).
 77-470999

Michaud, J. Fr. *Biographie universelle ancienne et moderne*. Nouvelle éd. Graz: Akademische Druck- u. Verlaganstalt, 1966-68.
CALL NUMBER: CT143 .M52 MRR Biog

   This second edition of a standard 19th century French biographical work (commenced in 1810) provides revised articles as well as immense augmentation in the number of biographies included. The biographies include political, scientific and literary luminaries of all times and all places.
 87-5040

D.   Immigration and Naturalization

The French in Texas : history, migration, culture. Francois Lagarde, editor. 1st ed.
Austin: University of Texas Press, 2003. xiii, 330 p., [16] p. of plates, ill. (some col.), col. maps.
CALL NUMBER: F395.F8 F74 2003
2002008289

Brasseaux, Carl A. **The "foreign French" : nineteenth-century French immigration into Louisiana**. Lafayette, La.: Center for Louisiana Studies, University of Southwestern Louisiana, c1990-<c1993 >. v. <1-3 >.
CALL NUMBER: F380.F8 B73 1990
   Incomplete Contents: v. 1. 1820-1839 -- v. 2. 1840-1848 -- v. 3. 1849-1852. Notes: Includes bibliographical references.
90081145

Votre nom en France (les naturalisations en France). 200p., deux couleurs, 40 illustrations, graphiques et cartes.
ON ORDER
   "...Fiches de naturalisation extraites du Journal officiel (1900 à 1960)...En un volume, ce guide pratique nous permet de retrouver l'histoire d'un aïeul, d'obtenir la copie de son dossier de naturalisation, de remonter le temps et de devenir, grâce aux nombreux classements, cartes et statistiques, un « généalogiste du monde » !" Review by publisher.

Bibliothèque nationale de France. Des sources pour l'histoire de l'immigration en France de 1830 à nos jours : guide. Sous la direction de Claude Collard. [Paris]: Bibliothèque nationale de France, 2006. 427 p.
CALL NUMBER: Z7164.I3 B55 2006
 2007370775

E. Military

Buffetaut, Yves. *Découvrir la carrière militaire d'un ancêtre*. Paris: Autrement, 2005. 88 p. : ill.
CALL NUMBER: CS596.S69 B84 2005
2005476561

Buffetaut, Yves. *Votre ancêtre dans la Grande Guerre: guide généalogique et historique*. Louviers: Ysec éditions, 2000. 255 p. ill.
CALL NUMBER: CS583 .B84 2000
2001363988

Bodinier, Gilbert. *Dictionnaire des officiers généraux de l'armée royale, 1763-1792*. Paris: Archives & culture, c2009- . v. <1>. Tome I, A-C.
CALL NUMBER: DC44.8 .B63 2009
   "Le Dictionnaire des officiers généraux de l'armée royale (1763-1792) de Gilbert Bodinier vient assurer le lien entre la Chronologie historique militaire de François-Joseph-Guillaume Pinard (1760-1778) et le Dictionnaire biographique des généraux et amiraux de l'Empire publié par Georges Six en 1934 et 1935. Il recense, en outre, les brigadiers de cavalerie et de dragons nommés après 1715, qui n'apparaissaient pas dans l'ouvrage de Pinard. Ce premier tome, qui couvre les lettres A à C, contient plus de 750 notices biographiques, précédées d'une bibliographie générale. Chacune d'entre elles présente, après le nom de l'officier, un bref rappel de l'origine de sa famille, la description de ses armoiries, puis la vie de l'individu : filiation, entrée dans la carrière, faits d'armes, mais aussi activités parallèles à la vie militaire, fortune et relations sociales. Les sources — principalement tirées du Service historique de la Défense, conservatoire des archives militaires françaises du XVIe siècle à nos jours —, et la bibliographie propre à chaque personnage apparaissent en conclusion, et permettent au lecteur curieux d'approfondir ses recherches.
   La richesse des notices de ce dictionnaire en fait une source inépuisable de renseignements, et le destine à devenir un instrument de travail indispensable pour l'étude de la société militaire dans les dernières décennies de l'Ancien Régime. Il intéressera aussi bien les amateurs d'histoire des familles que les chercheurs en histoire militaire, diplomatique, sociale ou institutionnelle." Review from publisher.
2009492168

Géhin, Gérard. *Dictionnaire des généraux et amiraux français de la grande guerre 1914-1918*.
Paris : Archives & culture, 2007-2008. 2 v. : ill., ports.
CALL NUMBER: D507 .G44 2007
   "... généraux ou d'amiraux de 1914 à 1918 ...ceux qui ont fait une carrière longue comme ceux qui se sont fait tuer peu après leur nomination (les généraux n'étaient pas obligatoirement des « planqués de l'arrière »), ils sont tous là, par ordre alphabétique, dans ce dictionnaire biographique : état civil, portrait, parcours, décorations, articles... Comme rien d'exhaustif n'existait sur les chefs militaires de la Grande Guerre, cette somme de travail comble une lacune importante. Résultat de dix années de recherches, elle constitue un ensemble de référence indispensable à toute bibliothèque comme à tout historien travaillant sur cette période."
Review from Publisher.

Tableau d'honneur de la Grande Guerre : planches. Rassemblées par Solange Contour; préface du colonel Rodier. Paris : Archives & culture, c2000-<2003> . v. <1-4>, ports.
CALL NUMBER: D609.F8 T33 2000 fol.
2001348430
"En deux tomes, les Éditions Archives & Culture avaient réédité les portraits des militaires décorés lors de la Première Guerre mondiale et publiés à cette époque dans L'Illustration. Le tome 3, paru en 2002, et le tome 4, en 2003, sont des œuvres originales, complétant les premiers tomes avec des militaires qui n'y figuraient pas. Les informations sont trouvées pour partie en archives et pour partie par les familles qui fournissent des portraits manquants. Le tome 5 est en préparation." Review from Publisher.

F. Land and Real Estate

Provence, Myriam. *Retracer l'histoire d'une maison.* Paris : Autrement, 2004. 80 p. : ill.
CALL NUMBER: DC36.9 .P76 2004
2005384932
How to trace a house history. Also useful for tracing land records.

G. Local History

Guide de l'histoire locale: faisons notre histoire! Sous la direction de Alain Croix et Didier Guyvarc'h. Paris : Seuil, 1990. 347 p.: ill., maps.
CALL NUMBER: D16.4.F7 G85 1990
Methodology of Local history. Includes bibliographical references and index.
91129317

Levron, Jacques. *L'histoire communale; esquisse d'un plan de travail. Éd. entièrement refondue.* Paris Éditions Gamma, 1972. 117 p. illus.
CALL NUMBER: DC36.9 .L43
Historiography and Local History of France. How to plan a local history of a commune.
74314343

Bottin communes. Paris: St. Didot-Bottin, 1978- .
CALL NUMBER: DC 14 .B63 Main Reading Room Alc. (1988)
List of regions in France. List of departments, their prefectures, and sub-prefectures. Arranged in regional order, each department has a map, list of administrative offices, and an alphabetical list of communes. Includes hamlets, remote or secluded places, named places, for metropolitan France and territories ("hameaux, ecarts, lieux-dits (France métropolitaine et outre-mer"). Each commune is briefly described (geography and administration). Alphabetical index of communes at back of book.
79-645925

Boyrie-Fénié, Bénédicte. *Dictionnaire des pays et provinces de France.* Bordeaux: Editions Sud Ouest, 2000. 349 p. : maps.
CALL NUMBER: DC14 .B68 2000
Describes the former provinces of France, which subdivided the country until the creation of

départements in 1789. Arranged alphabetically, each entry lists the names given to the area since Gallo-Roman times, facts about the area, and describes the geography of the area. An important source for the study of the history of France. Review: http://www.library.illinois.edu/mdx/french/freculture.html
 2001438508

Dauzat, Albert. *Dictionnaire étymologique des noms de lieux en France.* 1978
CALL NUMBER: DC14.D28 197
    Names of places in France.
79387605

Dauzat, Albert. *Dictionnaire étymolgique des noms de rivières et de montagnes en France.* 1978
GB1293 .D37 1978 EUR West FR (European Reading Room)
    Names of rivers and mountains in France.
 78390137

Dictionnaire des communes avec l'indication de la perception dont relève chaque commune. 1939
CALL NUMBER: DC14.D6 1939
    Dictionary of communes (towns) and a description of each in 1939. Another edition, with the call number DC14.D6 1946, provides descriptions of each in 1946.
4023201

Dictionnaire national des communes de France. Paris: A. Michel: Berger-Levrault, 1984.
CALL NUMBER: DC14.D63 1982 EUR West FR (European Reading Room)
    Dictionary arrangement of place names in France. Includes communes and principal towns, villages, bergs, hamlets, farms, wind and other mills, houses and country seats, parks, and other place names whose names have accumulated over hundreds of years.
bi94-15167

Gourdon de Genouillac, Henri. *Dictionnaire de fiefs, seigneuries, chatellenies, etc. de l'ancienne France contenant.* Paris: E. Dentu, 1862. 2 p. l., viii, 567 p.
CALL NUMBER: DC14.G6
    Dictionary of fiefs, seigneuries, and other historical small geographic designations.
 42049337

Histoire des provinces de France. Préface de Pierre Miquel. Paris: F. Nathan, 1981-1984.
8 v., ill. (some col.).
CALL NUMBER: DC38 .H513 1981
        Contents: 1. Bourgogne, Dauphiné, Savoie, Lyonnais -- 2. Alsace, Lorraine, Franche-Comté -- 3. Champagne, Ardennes, Flandres, Picardie -- 4. Normandie, Bretagne, Vendée, Maine-Anjou -- 5. Limousin, Poitou-Charentes, Aquitaine, Béarn, Pays basque -- 6. Paris, Ile-de-France, Touraine, Orléanais-Berry -- 7. Auvergne-Bourbonnais, Rouergue, Languedoc, Roussillon -- 8. Provence, Nice, Corse. Includes bibliographies.
 84121853

Pneu Michelin (Firm). *Dictionnaire des communes de France.* Paris: Pneu Michelin Services de tourisme, 1978.
CALL NUMBER: DC14 .M38 1978 Geography and Maps Reading Room

79356453

H. Dictionaries of Genealogical Terms

Genealogical word list. French. 1st ed. Salt Lake City, Utah (50 E. North Temple St., Salt Lake City 84150): Family History Library, Church of Jesus Christ of Latter-day Saints, 1990. 12 p. CALL NUMBER: CS6 .G46 1990
91135689

Michel-Gasse. *Dictionnaire-guide de la généalogie*. Paris: J.-P. Gisserot, 1999. 127 p.
CALL NUMBER: CS6 .M53 1999
99213067

Nemo, Alain. *Dictionnaire du généalogiste et du curieux*. Saint Marcel: A. Nemo, [1992]. 295 p.: ill., maps.
CALL NUMBER: CS6 .N46 1992
94173746

Roy, Léon. *Dictionnaire de généalogie* / préface de Michel Dorban. Bruxelles: Labor, 2001. 711 p.
CALL NUMBER: CS6 .R69 2001
2002385308

Tardif, H.-P. (Henri-P.). *Compléments de généalogie*. Sainte-Foy, Québec: H.-P. Tardif, [1999] 185 p.
CALL NUMBER: CS582 .T37 1999
    Discusses the connection between genealogy and other scientific disciplines. Contains a French genealogical vocabulary; and a list of genealogical quotes in French and English; and a French-English genealogical dictionary. Focus on examples from Quebec.
99491987

I. French Surnames

French Surname Meanings & Origins
Uncovering Your French Heritage
By Kimberly Powell, About.com
http://genealogy.about.com/cs/surname/a/french_surnames.htm
    "Coming from the medieval French word 'surnom' translating as "above-or-over name," surnames or descriptive names trace their use back to 11th century France, when it first became necessary to add a second name to distinguish between individuals with the same given name. The custom of using surnames did not become common for several centuries, however...."

Beaucarnot, Jean-Louis. *Les noms de famille et leurs secrets*. Paris: R. Laffont, 1988. 355 p.
CALL NUMBER: CS2695 .B394 1988
    French personal names and etymology. Bibliography: p. 305-307. Includes indexes.
88202741

Fordant, Laurent. ***Atlas des noms de famille en France***. Paris: Archives & culture, 1999. 190 p., ill., maps.
CALL NUMBER: CS2697 .F67 1999
    A French language resource which chronicles the geographical dispersion of surnames throughout France. It lists the most popular 1000 names and how they have changed rank in the last century. It also addresses the influx of non-French names into the country. Statistics are given for France as a whole, individual regions, and individual departments. Review from:
http://www.library.illinois.edu/mdx/french/freculture.html
99228610

Les noms de famille en France : histoires et anecdotes. Sous la direction de Marie-Odile Mergnac; préface de Jacques Dupâquier. 2. éd. augm. Paris: Archives & culture, 2000. 477 p., maps. CS2695 .N66 2000<LH&G>

    Names in dictionary order. Since it is imperative to know where a family was from in order to obtain records in France, one of the most useful features is a map showing the provinces in which the name is found, and the number of persons with this name in France today. Also given is the etymology of the name, as many as 20-30 histories of famous people by this name, stories connected to the name, variations (names) derived from this name.
2001351033

Fordant, Laurent. ***Tous les noms de famille de France et leur localisation en 1900***. Préface de Jean-Louis Beaucarnot. Paris : Archives & culture, 1999. 1420 p.
CALL NUMBER: CS2695 .F67 1999

    "Cette monumentale encyclopédie donne la fréquence et la localisation principale de tous les noms de famille présents en France au début du XXème siècle, de ceux qui s'y sont éteints comme de ceux qui y sont toujours représentés, même par une seule personne. Construit à partir du fichier des actes de naissance informatisé par l'Insee, avec plus de 600 000 patronymes cités, il constitue un ouvrage de référence essentiel pour toute bibliothèque et pour tout amateur ou association de généalogie. Il permet une bonne approche de la localisation ancienne des noms de famille puisque, en 1900, la mobilité des populations était réduite et que les enfants ne naissaient pas encore dans la maternité du chef-lieu. Un outil à consulter impérativement lorsque le nom est rare." Review from:
http://www.archivesetculture.fr/livre-archives-et-culture-noms-de-famille-et-de-lieux-tous-les-noms-de-famille-de-france-et-leur-localisation-en-1900-23.html
00356655

Les noms de famille du Sud-Ouest. Sylvie Monniotte, et al, avec la collaboration de Baptiste Levoir, et al. Paris: Archives & culture, [1999]. 294 p., ill.
CALL NUMBER: CS2699.F72 N66 1999
    Dictionary of personal names in Southwest France. Series: Collection Les noms de famille en France.
99510823

J.   Professions

Mergnac, Marie-Odile, *Les métiers de nos ancêtres.* Paris: Archives & culture, 2007. [Vendor Info: Jean Touzot Libraire Editeur (TOUZ) 29 EUR] 271 p., ill.
NOT AT LIBRARY OF CONGRESS
    Occupations of our ancestors. History of Occupations, "Old Jobs" in France. Series: Collection Vie d'autrefois; Variation: Collection Vie d'autrefois (Paris, France). Standard No: ISBN: 9782350770352; 2350770354. "Répertoire de métiers disparus ou ne se pratiquant plus de la même façon : coutelier, tisserand, dentellière, scieur de long, berger, terre-neuvas... Avec des pistes pour des recherches généalogiques en fonction des métiers exercés." pour tous les curieux ou tous les généalogistes qui veulent aller plus loin, des annexes en fin d'ouvrage indiquent, métier par métier, où et comment trouver des informations sur le lointain aïeul qui a pu l'exercer. Includes bibliographical references (p. 264-270).

www.ingramcontent.com/pod-product-compliance
Lightning Source LLC
Chambersburg PA
CBHW081824170526

45167CB00008B/3534